SEASONAL SCIENCE PROJECTS

Autumn

Science Projects

JOHN WILLIAMS

Julian Messner
Parsippany, New Jersey

First published in 1996 by Evans Brothers Limited
2A Portman Mansions
Chiltern Street
London W1M 1LE
England

© Evans Brothers Limited 1996

First published in 1998 in the United States by Julian Messner

A Division of Simon & Schuster
299 Jefferson Road
Parsippany, New Jersey 07054-0480

ISBN 0-382-39711-8 (LSB) 10 9 8 7 6 5 4 3 2 1
ISBN 0-382-39712-6 (PBK) 10 9 8 7 6 5 4 3 2 1

Acknowledgments
Planning and production by The Creative Publishing Company
Edited by Paul Humphrey
Designed by Christine Lawrie
Commissioned photography by Chris Fairclough
Illustrations by Jenny Mumford
The publishers would like to thank the staff and pupils of East Oxford First School for their help in the preparation of this book.

For permission to reproduce copyright material, the author and publishers gratefully acknowledge the following: Bruce Coleman (John Shaw) 6, (Hans Reinhard) 8, 10, (George McCarthy) 18; Oxford Scientific Films (Barrie E Watts) 17, (Stan Osolinski) 20, (Paul Franklin) 26 (left), (G I Bernard) 26 (right).

Library of Congress Cataloging-in-Publication Data

Williams, John, 1939 Jan. 11--
Autumn science projects/by John Williams
p. cm. (Seasonal science projects)
Originally published: London, England: Evans Brothers Ltd., 1996.
Includes index.
Summary: Presents a variety of projects and experiments appropriate to autumn, including studying the night sky, finding out how seeds are disperse, and making a model glider. Includes notes for parents and teachers.
1. Autumn--Experiments--Juvenile literature. 2. Science projects--Juvenile literature. [1. Science projects. 2. Science--Experiments. 3. Experiments. 4. Autumn.] I. Title. II. Series: Williams, John, 1936-Seasonal science projects.
QB637.7.W55 1996   96-20182
507'.8--dc20   CIP
                                          AC

Contents

* Words in bold in the text are explained in the glossary.

What is autumn?

There are four seasons—spring, summer, autumn, and winter. In autumn the air begins to get cooler. Animals and plants start to get ready for winter.

Here are two signs that autumn has come:
■ The leaves on some trees begin to turn red, orange, or brown.
■ It begins to get dark earlier in the evenings.
Can you think of any more?

Autumn is also called the "fall." Why do you think this name is used?

Autumn is not the same everywhere. In some countries it can be a very short season. You can find out which countries have long autumns and which ones have short autumns.

Autumn is the time of year when the leaves on some trees start to turn red, orange, or brown.

Project: Autumn around the world

You will need
- A world map or globe
- A notebook and pencil

What to do

1. Find the **lines of latitude** called the Tropic of Cancer and the Tropic of Capricorn on your map or globe.

The countries between these two lines have warm weather almost all the year round. This region is called the tropics.

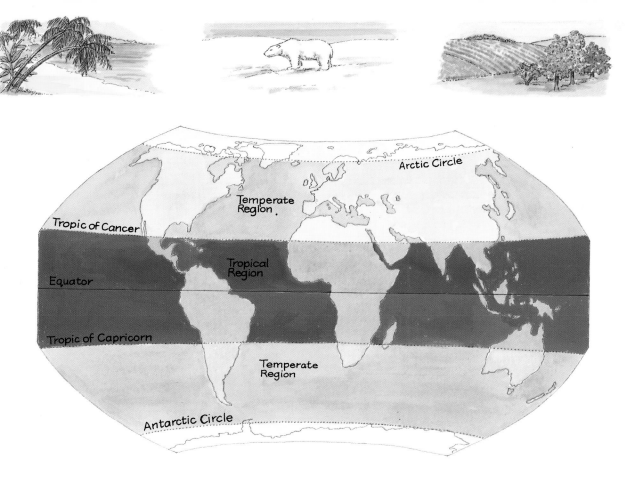

Arctic Circle

Temperate Region

Tropic of Cancer

Tropical Region

Equator

Tropic of Capricorn

Temperate Region

Antarctic Circle

2. Now look for the lines of latitude called the **Arctic** Circle and the **Antarctic** Circle.

The countries to the north of the Arctic Circle and south of the Antarctic Circle have little or no autumn. They have short summers and long winters.

3. List some of the countries between the Arctic Circle and the Tropic of Cancer and some of those between the Tropic of Capricorn and the Antarctic Circle. (Where does the United States land?)

These are the temperate countries, where there will be a long autumn.

Autumn fruits

Many different kinds of fruits are picked in the autumn. The fruit is the part of a plant that carries the seeds.

Harvesting apples in autumn.

There are many different types of fruits. Here are three types that you might find:

Berries are soft fruits that often have many little seeds covered by a thin skin. Tomatoes and gooseberries are berries.

Pods are long, narrow fruits, with a row of seeds inside. Peas and runner beans grow in pods.

Drupes are soft fruits with a hard pit inside. The pit has the seed inside it. There is only one seed in each pit. Cherries and plums are drupes.

Project: Collecting fruits

1. Make a collection of as many different fruits as you can find. Look in orchards and your garden as well as the supermarket.

2. Divide your collection into berries, pods, and drupes.

SOME FRUITS ARE POISONOUS. MAKE SURE YOU HAVE AN ADULT WITH YOU WHEN YOU MAKE YOUR COLLECTION.

3. Ask an adult to help you cut the fruits open. Can you find the seeds? How many seeds are there? Are they big or small, hard or soft?

4. Make drawings of the cut halves of fruits.

Project: Making fruit prints

You will need
- Different kinds of hard fruits, like apples and pears
- Paper
- A knife
- Paint and a paintbrush
- Saucers to mix paint in

What to do

1. Ask an adult to cut the fruits in half down the middle.

2. Cover the cut half with paint and press it down onto the paper. Do this all over your paper to make a pattern. Try using different colored paints.

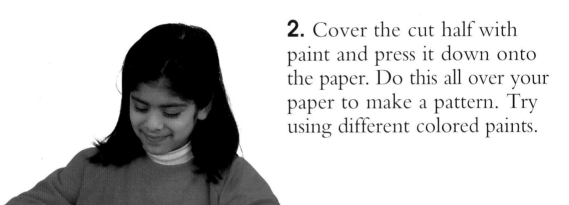

Fruit stains

If you have been out picking blackberries, you will find your fingers are colored blue. This is the juice of the berry. If it stains your clothes, it is very difficult to remove.

Blackberries are a sure sign of autumn and are a delicious feast for people, animals, and birds.

Project: Removing fruit stains

You will need
- Some pieces of old white cotton sheet, all about 8 in x 8 in
- Plastic pots or cups
- Warm water
- Juice squeezed from blackberries, peaches, or some other soft fruit
- Some different kinds of detergent powder
- A teaspoon

What to do

1. Put one drop of fruit juice in the middle of each piece of cloth.

2. Pour equal amounts of warm water into each cup.

3. Stir a small teaspoonful of the different powders into each of the cups but leave one cup with no detergent in it at all. Label the cups carefully.

4. Put a cotton square into each cup. Stir for ten minutes.

5. Take out the pieces of cloth and dry them.

Have any of the stains faded?

Have any of them disappeared altogether? What happened to the cloth in the cup with no detergent powder?

Arrange the cloths in order from most to least faded.

Are fruits acid?

Many fruits can taste sour. This may be because they are not ripe. But some fruits, such as lemons, limes, and even some oranges are always sour. The sourness is caused by chemicals in the fruit called acids.

Project: Testing for acids

You will need
- Some clear plastic cups
- Some fruit. Try a lemon, an orange, and blackberries
- Litmus paper

What to do

1. Squeeze a little fruit juice into each of the containers, one juice for each cup.

2. Dip a piece of blue **litmus** paper into each of the cups.

If the paper turns red, the juice is an acid. If the paper stays the same color, the juice is not.

Did all the juices turn the litmus paper red?

Project: Make your own acid indicator

You will need

- Red cabbage leaves or dahlia petals
- A bowl
- A short piece of smooth stick
- A little water
- A plastic cup

What to do

1. Crush the red cabbage leaves or dahlia petals with a little water, using the piece of smooth stick.

2. Collect the liquid in the plastic cup. This is your acid **indicator**.

What color is the liquid?

3. Pour a little of the liquid into each of your fruit-juice samples.

Do the samples change color?

Compare the colors that your homemade indicator gives compared to the litmus paper.

Can you tell if the fruit juices are acid, using your homemade indicator?

Can you tell how strong the acid is?

How seeds spread

When seeds are ripe, they must be planted. However, if the seeds just fell all together around the parent plant, they would not be able to grow well. They will grow better if they are spread around, or dispersed. This is done in many ways.

Look at your fruit collection. How do you think the seeds inside them are dispersed?

Some trees, such as the ash, plane, sycamore, and maple, have seeds with wings. These fly like helicopters and land many feet from the tree.

You can experiment with flying seeds.

Project: Flying seeds

You will need
- Some different kinds of seeds with wings
- A chair
- A long tape measure

What to do

1. Take your seeds out into the playground or garden. Stand on the chair and throw them into the air, one at a time. How do they fly?

2. Use your tape measure to measure how far each seed flies. You can record your results on a bar graph like the one below.

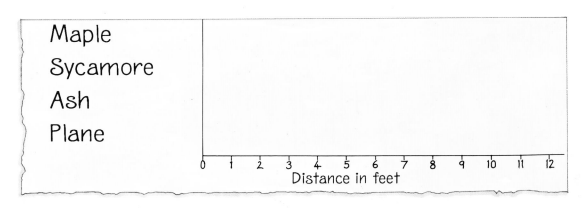

Maple
Sycamore
Ash
Plane

0 1 2 3 4 5 6 7 8 9 10 11 12
Distance in feet

Project: Make your own sycamore seed

You will need
- Some poster paper about 6 in x 2 in
- Modeling clay
- A straw
- Glue
- A chair

What to do

1. Cut the paper into the shape of a sycamore wing.

2. Bend the straw into the shape of one edge of the wing and glue it to that edge.

3. Glue the modeling clay to the end of the wing to form the seed. Twist the wing into the shape of a propeller.

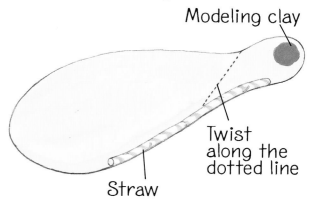

Modeling clay

Twist along the dotted line

Straw

4. Stand on a chair and drop the seed.

You can make several models of different sizes. Which one flies the best?

Plants without seeds

Mushrooms and toadstools belong to a group of plants called fungi. Fungi flourish in autumn because there is plenty of rain and it is not too cold. Above all, there are plenty of dead leaves for them to grow on. Fungi get their food from dead and rotting plants.

Look at the underside of a mushroom. You will see lots of gills. These are where the spores are made. Spores are like seeds, and the mushroom has many hundreds of them.

Project: Making spore patterns

You will need
- Some large mushrooms with the stalks cut off
- A sheet of white paper

What to do

1. Place the mushrooms with the gills down on the paper.

2. Leave them like this for several hours, then gently remove the mushrooms.

Can you see the patterns left behind? These are made by the spores that have fallen out of the gills.

WARNING: MANY FUNGI ARE POISONOUS. NEVER PICK THEM FROM THE WILD.

Right: Not all fungi simply drop their spores. Some, like the puffball, send them shooting into the air. There they fly away on the breeze to settle and grow elsewhere.

Did you know that yeast is a kind of fungus? Yeast is used to make bread. It produces a gas, which makes the bread rise. Yeast is also used in the making of beer and wine.

Project: Watching yeast at work

You will need
- A small plastic bag
- Sealing tape
- Some dried yeast
- Water
- A little sugar

What to do

1. Make up a mixture of yeast and water, following the directions on the packet. You may need to add sugar.

2. Pour the mixture into the plastic bag.

3. Seal the bag with the sealing tape to make sure that no air can get into it.

4. Put the bag in a warm room away from drafts.

What happens to the yeast mixture? What happens to the plastic bag?

Birds and flight

It is not only seeds and spores that fly away in autumn. Birds like swallows and martins fly great distances to warmer countries for the winter.

Birds, like this blue tit, are the masters of the air. Their bodies are specially designed for flight, with muscles that are very powerful for their size.

To fly well, a bird must have a strong but light skeleton. It must have large lungs to take in extra **oxygen** for the hard work of flying. And, of course, it must have wings with feathers.

There are two main sorts of feathers. The soft down feathers keep the birds warm by trapping air. The larger contour feathers have several uses. The most important contour feathers are those on the wings. These are called flight feathers.

Project: Make a feather helicopter

You will need
- Four long flight feathers
- A piece of dowel rod about 8 in long
- A cork with a hole through the middle
- Modeling clay
- A stopwatch
- A chair

What to do

1. Push the feathers into the cork around its edge, making sure that they are evenly spaced.

2. Push the dowel rod through the middle of the cork until about ⅜ of an inch shows at the top.

3. Twist the feathers so that they are at an angle to the cork, like the blades of a propeller.

4. Fix a ball of modeling clay to the long end of the dowel rod.

5. Stand on a chair and drop your helicopter. Measure the time it takes to reach the ground with your stopwatch.

You may need to adjust the angle of the feathers and the amount of clay you use to make the helicopter work properly.

Looking at feathers

All birds have feathers, even birds that don't fly. The central part of the feather is the shaft. Out of the shaft grow hundreds of **barbs**. These are locked together by millions of tiny hooks.

When all the wing feathers are overlapped, they make a strong, flexible wing.

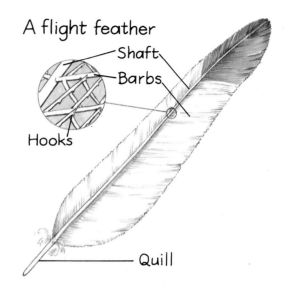

A flight feather

Shaft

Barbs

Hooks

Quill

Project: Looking at feathers

You will need
- A collection of wing and downy feathers from different birds
- A bowl of water
- A magnifying glass

What to do

1. Take one of the feathers and run your finger down the barbs toward the quill. Can you separate the different barbs?

2. Run your finger back up toward the tip of the feather. What happens to the barbs you have separated?

3. Dip the feather in water. Take it out and shake it. Does it stay wet?

Left: Birds must take great care of their feathers. Removing dirt and dust is called preening.

4. Repeat these steps with the other wing feathers. Are they all the same?

5. Now repeat these steps with the downy feathers. What happens?

6. Examine a wing feather under a magnifying glass. Can you see the tiny rows of hooks along the barbs?

Gliders

 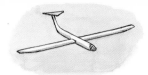

Birds flap their wings so that they can move around in the sky and to take off and land. Once they are in the air, some large birds use their wings to glide. Albatrosses can glide for days without ever flapping their wings.

You can make a glider of your own and test how far it can fly.

Project: Make a glider

You will need
- A long, thin tube of light plastic foam about 32 in long (the kind that is used to insulate water pipes)
- Three sheets balsa wood $\frac{1}{8}$ in thick: one 4 in x 35 in, one 4 in x 12 in and one 4 in x 5 in
- Rubber bands
- Modeling clay
- Two pieces of an index card, each about 3 in x 1 in
- Glue
- Sandpaper

What to do

1. Round off the ends of the three pieces of balsa wood with the sandpaper.

2. Fix the long balsa wood sheet about a third of the way along the top of the foam tube, using the rubber bands. This will form the wings.

3. Fix the second piece of balsa wood near the back of the tube at the bottom. This will form the tail.

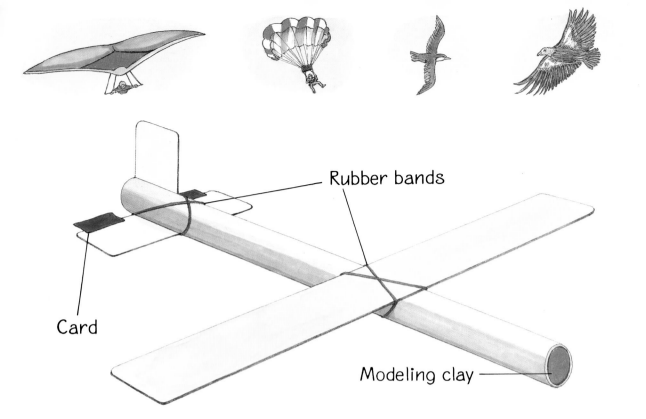

Rubber bands

Card

Modeling clay

4. Slide the last piece of balsa wood into a slit in the tube. This will form the **rudder**.

5. Glue a piece of card to each back edge of the tail. These will help the glider to fly.

6. Hold the glider carefully by the wings. Add clay to its nose until it balances.

7. Test your glider to see how far it can fly.

Real gliders are towed into the air by small airplanes and glide on updrafts of warm air.

WARNING: ASK AN ADULT TO HELP YOU CUT THE BALSA WOOD.

The night sky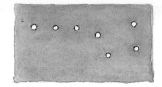

Autumn is a good time to look at the night sky. It gets dark earlier in the evening, so you can look at the stars before you go to bed.

Astronomers can recognize the hundreds of stars by their position in the sky. Positions can change during the year. They also differ when seen from different parts of the earth.

One of the best ways to identify stars is to find the ones that make up groups called **constellations**.

Look at the star charts on these pages. They show some of the constellations that you can see from the northern and southern parts of the earth. Can you recognize any of these constellations in the night sky where you live?

A star tube will help you identify the constellations.

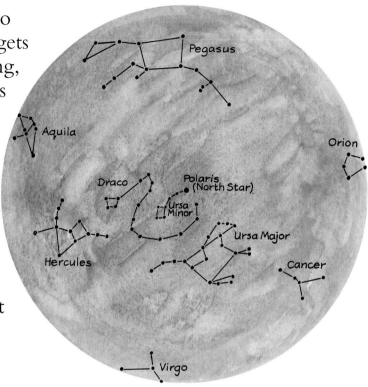

The northern sky.

Project: Making a star tube

You will need
- A large cardboard tube
- Some round pieces of black paper, each a little bigger than the end of the tube
- A white pencil
- Rubber bands
- A pin

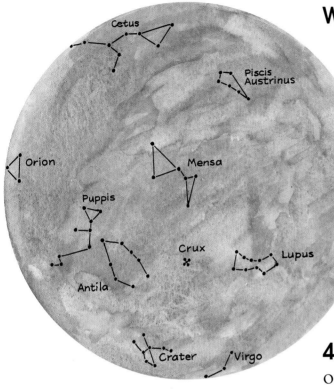

The southern sky.

What to do

1. Draw a circle around the end of the tube onto a piece of black paper.

2. Make pinholes within this circle in the shape of one of the constellations.

3. Cut some slits around the edge of the big circle.

4. Fix the black paper circle over the end of the tube with a rubber band.

5. Look down the other end of your tube to see the stars.

6. Mark different constellations on the other pieces of paper.

Animals in autumn

Some animals use the autumn to prepare for their winter hibernation. Others prefer basking in the autumn sun.

The red admiral butterfly (below left) makes the most of the autumn sun, while the beetle (below) prefers damp shade.

Here are three experiments to see what kind of **environment** wood lice prefer.

Project: Observing wood lice

You will need
- 10 wood lice
- A deep tray
- A piece of black paper
- A table lamp
- Some damp sand

What to do

1. Cover half of the tray with the black paper and shine the light onto the other half.

2. Put the wood lice onto the tray. Watch how they move around.

3. Count the wood lice in each part of the tray every minute for 10 minutes.

4. Make a record of your count as a graph like this.

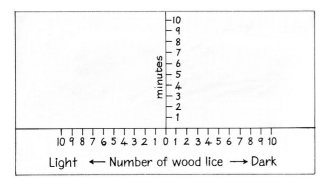

5. Now repeat this experiment, but this time place half of the tray in a warm room away from drafts. Cover the whole tray with the black paper.

6. Remove the paper every minute for 10 minutes and record where the wood lice are.

7. Repeat steps 5 and 6, but this time fill half of the tray with damp sand.

8. Return the wood lice to the place where you found them.

Put all your results onto a chart like this:

Environment	Number of wood lice after 10 minutes
Light	
Dark	
Dry	

Which of these environments do the wood lice prefer? Which of these environments would you prefer?

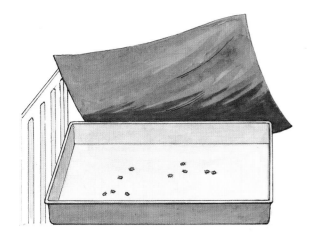

> **REMEMBER TO TREAT THE WOOD LICE WITH CARE. THEY ARE LIVING ANIMALS.**

Notes for Teachers and Parents

What is autumn? (Pages 6-7)
The seasons are primarily controlled by two factors: the position of the earth in its orbit around the sun, and the inclination of the earth's axis. When the Northern and Southern Hemispheres are pointed toward the sun, it is summer in the respective hemispheres. When the sun is directly overhead the equator, it is autumn and spring. Other seasonal variations can be explained by the sun's relative position to the earth. In summer the sun is higher in the sky, and the sun's rays are concentrated in a small area. In winter the rays hit that part of the earth at an oblique angle and are more spread out. Autumn can be thought of as a transitional period between summer and winter.

Autumn fruits (Pages 8-9)
Fruits are the ripened ovary of a plant, in which the seeds are formed. Most fruits ripen following pollination of a flower. There are some fruits that develop normally, but for some genetic reason have no seeds. Bananas are an example of such a fruit. Trying to classify fruits can lead children into difficulty. A blackberry, for example, is not a berry at all, but a collection of little drupes. Many nuts are in fact drupe pits. The hazelnut and the sweet chestnut are true nuts, but the walnut and the almond are drupe pits. The rose hip, because it is formed from different parts of the flower, is often not considered to be a true fruit at all. Common sense suggests that it should be a drupe.
An important botanical differentiation is made between biologically dry fruits, such as the poppy capsule, and the pods of legumes and those fruits that remain succulent. Virtually all other fruits fall into this category.

Fruit stains (Pages 10-11)
The activities in this section are an example of a controlled fair test. All the conditions should remain constant—the amount of water and its temperature, the amount of detergent, the amount of stirring, and the stains themselves should be constant. The water need not be too hot. Nearly all the main brands of powder or liquid detergent can be used within a wide variation of temperatures starting at 86°F. Care should be exercised if enzyme-action powders are used because some children may be allergic to them.

Fruit acidity (Pages 12-13)
This section introduces children to the elementary chemistry of some common substances. At this stage only acids are recognized, all other substances being nonacids. Children can later be introduced to the fact that the "opposite" of an acid is an alkali. They will need to understand that strong alkalis are as unpleasant as acids, and therefore great care has to be taken when using them. Simple litmus paper, available from most good drugstores, need only be used at this stage. Blue litmus paper will be turned red by all acids. There are other indicators that give different colors for different concentrations of acid or alkali. The best one to use to give a complete range of colors is universal indicator. This changes color to a deep red for a strong acid (pH 0-1) to deep violet for a strong alkali (pH 13-14). The midpoint of the scale shows a change from yellow to green at pH 7. Indicators made from red cabbage and other plants will also give a wide range of colors. These will not, however, all be the same, so a comparison with a known indicator is essential before any conclusions can be drawn.

Seed dispersal (Pages 14-15)
Most children understand the planting of seeds. This section asks them to consider plant distribution from the starting point of seed dispersal. Although most plants, particularly the larger ones, only seed in the autumn, many, the dandelion for example, seed several times a year. The dandelion is well-known for its form of seed dispersal. Other methods of dispersal include those employed by fruits such as the burdock, which have hooks that can catch onto an animal's fur (or a human's clothes). Many succulent fruits are eaten by animals, and the seeds are subsequently dispersed in the animal's droppings. Many pod fruits split open when they dry, showering their seeds in all directions. Some tropical fruits like coconuts can be dispersed for many miles simply by falling into the sea.
The work suggested with the well-known winged seeds will not only help children to understand this important process, but also introduce them to a more scientific approach to basic ecology.

Plants without seeds (Pages 16-17)
Fungi are a special group of plants. They have a simple structure and are dependent upon the absorption of organic materials for their food. They have no chlorophyll, but in other regards are a plant. What we see of

the fungi are the reproductive structures; most of the remaining plant is inconspicuous. The well-known fungi are particularly visible during the early part of a wet and warm autumn. They produce thousands of spores and have very efficient dispersal mechanisms. Fungi spores differ from seeds in that they have no embryo.

There are many thousands of different species of fungi, and yeast is one of the most simple forms. It is made up of a single cell and reproduces by budding. Various forms have many industrial uses. These strains, all originating from the original wild yeast, *Saccharomyces cerevisiae*, include brewer's and baker's yeast, as well as those used for the making of wine. During the fermentation process sugar is converted to alcohol, and carbon dioxide gas is produced. This is the gas that expands the plastic bag, and raises the bread.

Birds and flight (Pages 18-19)

It is very difficult to show children how birds use their wings to fly. It is probably best that they should see a videotape. Making the helicopter will at least give them hands-on experience of using the feathers in flight, albeit in a limited fashion. It is very important to make sure that the feathers are at the correct angle, or pitch. Children will have to experiment with this to make the helicopter spin.

Looking at feathers (Pages 20-21)

Many bird topics stop short at a study of the bird's anatomy. However, they are the only animals to be equipped with feathers, so a close look at these structures is valuable. The flight, or contour, feathers cover the body, and include the primaries, secondaries, and the tail feathers. A typical flight feather has a central shaft, the rachis. The quill is the thick end of the shaft. Attached to this is the vane, or web, of barbs. When the barbs of the web are separated, minute hooks can just be seen. These are the barbules, and on these are even smaller barbicels, which act like fasteners. In this way the web can be reformed when the bird is preening.

The down feathers are the fluffy feathers found on nestlings, but in many birds they persist through life. They keep the bird warm. A third very simple structure is the filo plume. These can usually be seen only when the other feathers have been removed.

Gliders (Pages 22-23)

The study of flight involves a knowledge of forces. Whatever propels the object forward is called thrust. The forward movement produces lift. This is brought about by the movement of air over the wings, so forming a partial vacuum. The air pressure is therefore higher on the underside of the wings, forcing it upward. This is the so-called "Bernoulli principle": the faster the flow the lower the pressure. There are two more forces at work on any flying machine: drag, the opposite to thrust, is the friction on the machine caused by its passage through the air; and gravity, which will try to pull it down.

The technology of making gliders is simple, although there are some important things to note. You should be able to balance the glider by placing your fingertips under the ends of the wings. You can make the glider balance by a combination of weight in the nose, and the position of the wings along the body. Tail flaps (elevators) will make the glider go up or down, and when flat, provide extra lift.

Children can experiment with different wing configurations, with the tail at the front and the wings at the back, for example.

The night sky (Pages 24-25)

Astronomy is a difficult subject to study at school because it is a nighttime activity. However, children can make these simple star tubes during a lesson about space. The technology needed to make them is basic, but tracing the constellations will introduce them to another aspect of the universe. Stars vary in magnitude (brightness) even within constellations. This can be shown by making a slightly bigger hole for the brighter stars.

Animals in autumn (Pages 26-27)

This section allows children to carry out some simple ecological and behavioral experiments. Most of the studies they make at this level are of the basic observational kind. This work will help children understand the habits of animals as well as their ecological needs. It will also help them understand science and experimentation.

The wood louse is not an insect as its name seems to suggest, but a terrestrial crustacean—the same group that includes shrimps, crabs, and lobsters. Choice chambers can be made out of two separate containers with only a small corridor between them. However, the tray suggested for this experiment is quite sufficient. Children should be taught to handle the creatures with care and respect, and not to subject them to extremes of heat or cold.

Glossary

Antarctic The part of the world around the South Pole, where the land is always frozen and covered with snow.

Arctic The part of the world around the North Pole, where the sea is always frozen.

Barb The jagged vanes on a bird's feather.

Constellation A small group of stars that appear to stay the same distance from each other in the night sky. They are often given a name.

Environment The place where an animal or plant lives, and all the many things that affect its life.

Indicator In science this is a substance that changes color when put into acids or alkalis.

Lines of latitude Imaginary lines drawn around the earth.

Litmus paper An indicator made from a plant dye.

Oxygen The gas in the air that we and almost all other living things need to live and breathe.

Rudder The the tail of an airplane, or part of a boat, needed to steer the craft.

Index